Corporate Security Organizational Structure, Cost of Services and Staffing Benchmark

Corporate Security Organizational Structure, Cost of Services and Staffing Benchmark

Research Report

Bob Hayes
Greg Kane
Kathleen Kotwica

ELSEVIER
AMSTERDAM • BOSTON • HEIDELBERG • LONDON
NEW YORK • OXFORD • PARIS • SAN DIEGO
SAN FRANCISCO • SINGAPORE • SYDNEY • TOKYO
Security Executive Council

Elsevier

The Boulevard, Langford Lane, Kidlington, Oxford, OX5 1GB, UK
225 Wyman Street, Waltham, MA 02451, USA

First published 2013

British Library Cataloguing-in-Publication Data
A catalogue record for this book is available from the British Library

Library of Congress Cataloging-in-Publication Data
A catalog record for this book is available from the Library of Congress

ISBN: 978-0-12-417002-5

This book has been manufactured using Print On Demand technology. Each copy is produced to order and is limited to black ink. The online version of this book will show color figures where appropriate.

Printed and bound by CPI Group (UK) Ltd, Croydon, CR0 4YY

Transferred to digital print 2012

CONTENTS

EXECUTIVE SUMMARY

This research report summarizes the results of a large security survey conducted by the Security Leadership Research Institute (SLRI) between October 2009 and June 2011. Researchers set out to benchmark the state of the security industry in terms of organizational structure, budget, staff resources, board-level risk concerns, program drivers, services provided, and more. The decision about what elements to include in this work was informed by extensive input from experienced security practitioners and was focused on providing the metrics they would find most useful. The collected data can be used by security professionals for introspective analysis of the security team within their organizations, external review of the ways in which other organizations approach risk management, internal justification of security budgets and initiatives, and for performance metrics.

The 183 survey respondents were risk management practitioners comprising a broad cross-section of industries. The organizations represented come from the public and private sectors and range from small- and medium-sized businesses to some of the largest international corporations. When possible, results are broken down into categories so readers can select specific metrics for comparison purposes. These details are provided when there is sufficient data to compare between categories or groups, and when anonymity is assured. Not all questions were answered by all participants.

WHAT IS A RESEARCH REPORT?

A research report provides the foundational information security practitioners need in order to review, understand, and successfully address key issues within the workplace. It presents current data collected and analyzed as a result of the Security Executive Council's Security Leadership Research Institute (SLRI) research. This research report can be used by security professionals and managers who want to identify successful practices of their peers in the field to prepare a business case, develop a strategy, or incorporate findings for their security programs and services that mitigate risk. Educators can use this report for assignments, class projects, and scenario development and analysis.

Survey Results

SECTION I: ORGANIZATIONAL DEMOGRAPHICS

In this section, information about the participants in this survey is provided. These demographics are used later in the report to provide metrics that enable organizations to compare apples to apples. A key takeaway from this first section is the diversity of the respondents' organizations.

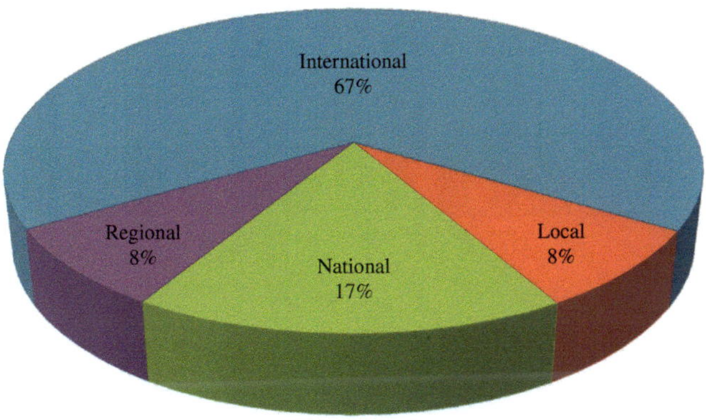

Figure 1.01 The geographic scope of the organization's business (note: the original survey question defined regional as regional within a nation).

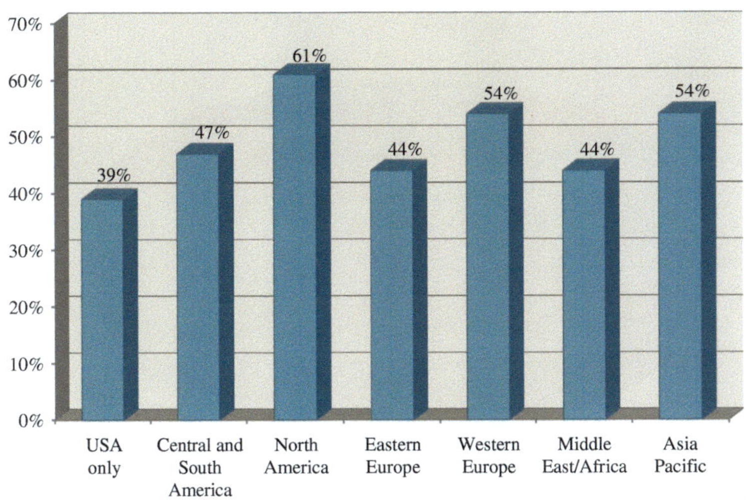

Figure 1.02 The regions of the world the organizations operate in (respondents could choose multiple regions).

Although a large number of industries are represented (see Figure 1.03), there were too few participants in this survey to warrant categorizing all of the following results by industry. Instead of providing possibly misleading metrics, it was decided not to provide a breakdown by industry for most of the analysis that follows.

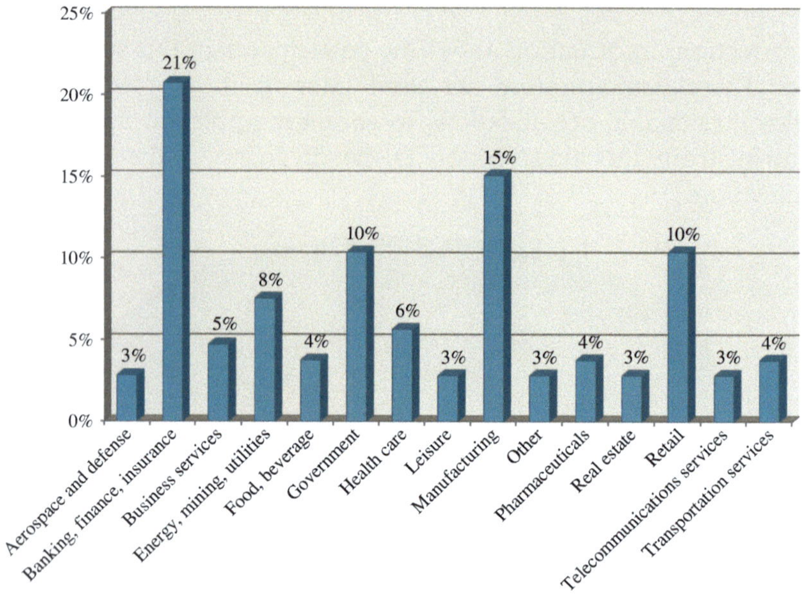

Figure 1.03 The industries represented by the respondents' organizations.

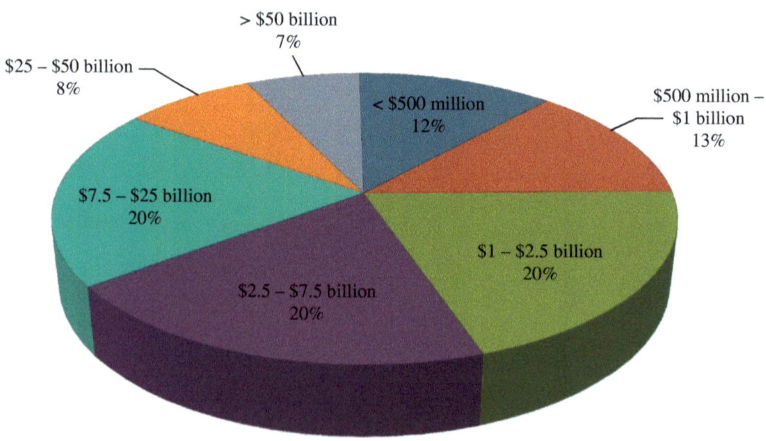

Figure 1.04 The organizations' annual revenue (revenue was converted to U.S. dollars when appropriate).

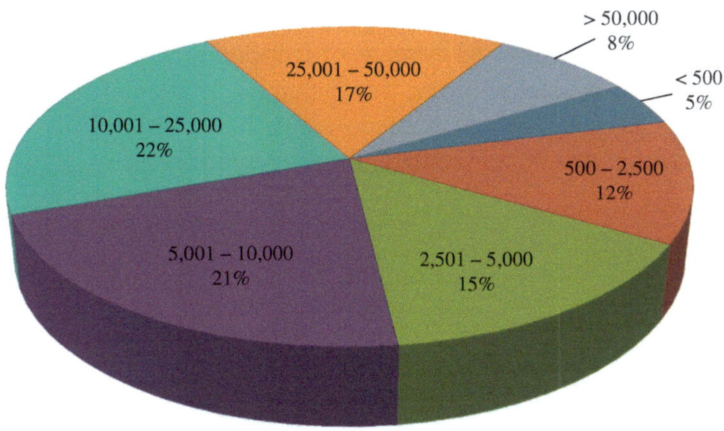

Figure 1.05 The total number of employees per organization.

SECTION II: RESPONDENT DEMOGRAPHICS

In this section, the typical characteristics of the practitioners participating in the survey are examined.

Due to the many different titles used in the security and risk management field, the survey asked the participant to select the title level he or she felt best matched his or her position in the organization (see Figure 2.01).

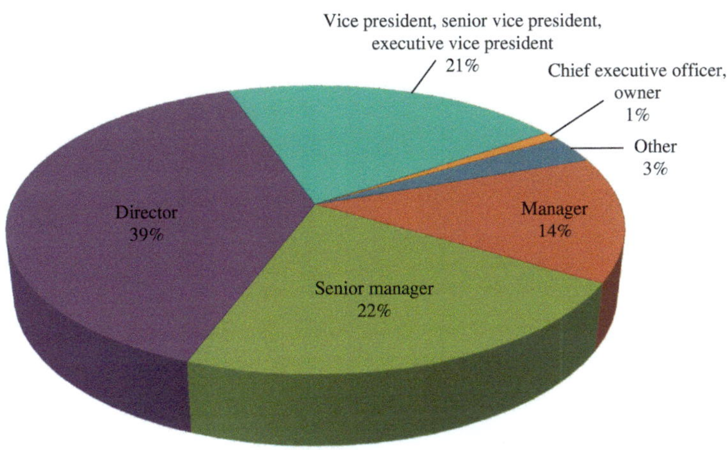

Figure 2.01 The title level of individuals participating in the survey.

Further analysis of the data collected about security areas managed (Figure 2.02) showed that 53 percent of respondents managed only corporate security/physical security, and 7 percent managed only information security. Sixteen percent of respondents had responsibility for both corporate security/physical security and information security.

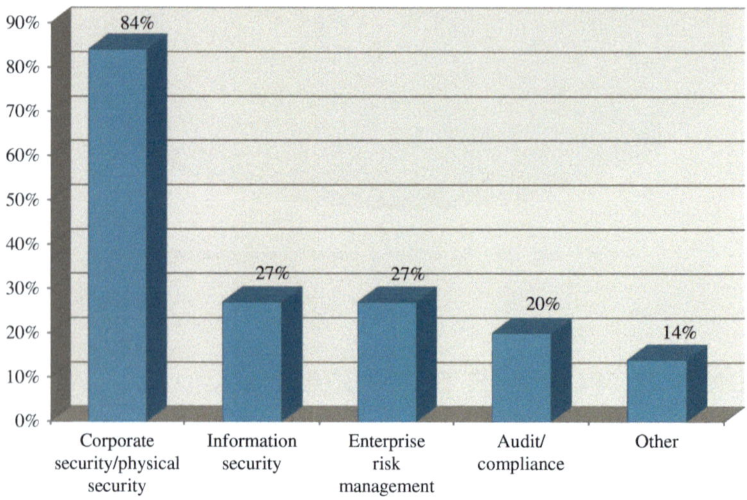

Figure 2.02 The security areas managed by respondents (respondents could choose multiple areas).

To get a better idea of the level within the organization in which the typical security and risk management executive operated, two questions were asked: what is the title level that best represented who they reported in to (Figure 2.03), and how many reporting levels the respondent is removed

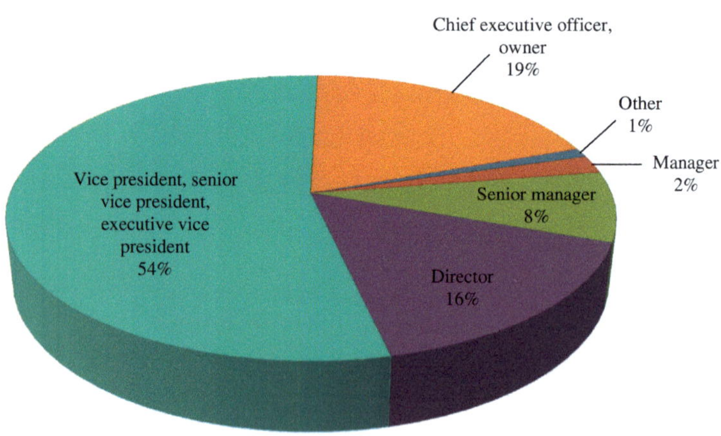

Figure 2.03 The reporting level of respondents within their respective organizations.

from the organization's senior-most operating executive (Figure 2.04). The data in these two figures show that 73 percent of respondents report to a vice-president level individual or above, and 61 percent are within two reporting levels of the senior-most operating executive.

Further analysis of the levels of reporting indicated that the size of the organization did not appear to be a determining factor in whether or not a security executive is likely to directly report to the senior-most operating executive.

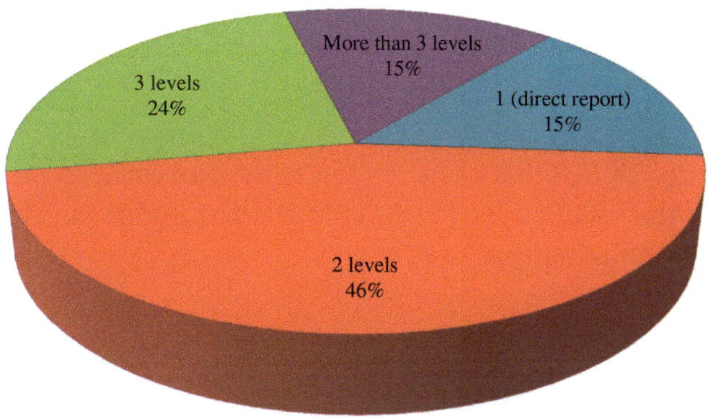

Figure 2.04 The number of levels of reporting between the respondent and the organization's senior-most operating executive.

There was little commonality regarding the business area in to which the security groups report (Figure 2.05), with no clear correlation between the business area and any attribute measured in this survey,

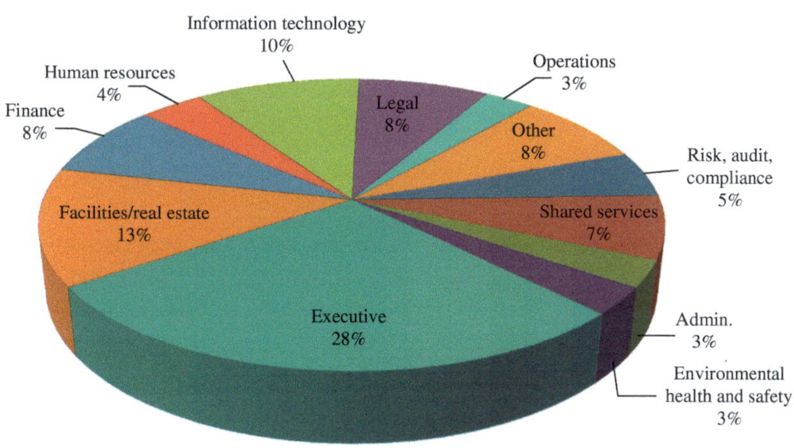

Figure 2.05 The business function to which each respondent's security group reports.

including industry, organization size (measured by revenue and employee count), scope of organization, and scope of responsibility.

Examples of functional areas that fell into the "other" category consisted of:

- Security/corporate security
- Asset protection
- Director of police and security/law enforcement
- Security, crisis management, business continuity
- Global physical security
- Loss prevention
- Supply chain

Another distinction between respondents was their scope of responsibility. For the majority of respondents, the scope of responsibility spanned multiple countries.

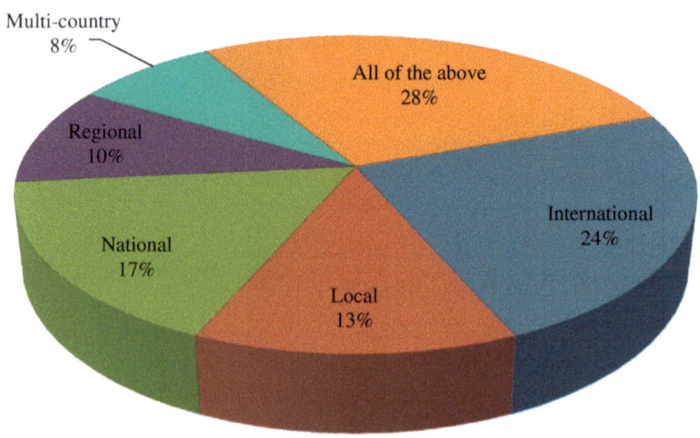

Figure 2.06 The scope of the respondent's responsibility.

In the following table, the respondent's scope of responsibility was compared to organizational scope of responsibility. In an apparent contradiction, some respondents stated that they have international/global scope of responsibility but work for a national organization. This could be attributable to the fact that the respondent simply didn't mean to select the answer they did. However, it is also likely that some respondents *do* need to address risk globally for a national organization; for

A breakdown (by percentage) of the business scope of the organizations in which the respondents who specified a scope of responsibility are employed. For example, of the respondents that stated they have local city or metropolitan area responsibility, 28 percent work in a global organization, 61 percent work in a local organization, 6 percent work in a national organization, and 6 percent work in a regional organization.

Respondent's Scope of Responsibility	The Organization's Business Scope			
	Global	Local	National	Regional (within a Nation)
All of the following	89	0	5	5
International/global	97	0	3	0
Local city or metropolitan area	28	61	6	6
National	30	0	70	0
Regional (portion of a nation)	29	0	21	50
Regional (consisting of multiple countries)	100	0	0	0

example, they may need to deal with risk issues with overseas suppliers even though the company only markets and sells nationally.

SECTION III: SECURITY BUDGET

Every organization is different, and subsequently security budget allocation varies widely from company to company. Survey respondents were asked to provide both their annual security budgets as well as their organization's annual revenue. For government entities, appropriated budgets, as provided in their latest available budget reports, were used in place of annual revenue.

Included below are averages across several different categories, including number of security employees, number of organization employees, organization revenue, organization scope, industry, and more.* The purpose of this breakdown is to provide risk management professionals with a way to obtain a rough estimate of typical security budget size. All monetary figures are in U.S. dollars.

The overall averages may appear to be reasonable for typical security budgets. However, caution should be exercised when interpreting

*Analysis of the security budget broken down by the number of reporting levels from the senior-most executive was inconclusive. Most likely this is related to the finding that the organization's size was not a determining factor in whether or not the respondent directly reported to the senior-most operating executive in the organization.

any further breakdown of those results, as the number of participants in this survey is too small to obtain a high degree of confidence in some of the categories. Notable findings to emerge from this analysis (and are included in the tables below) are:

- The overall average security budget was $8 million.
- The overall average security budget per organization employee was $249.
- The overall average security budget as a percentage of organization revenue was 0.07 percent.

The average security budget, broken down by the size of the security group within the organization (by employee count). An average for all respondents was also included.

Number of Security Employees	Average Security Budget
<10	$886,067
10–50	$2,627,905
51–100	$3,856,940
101–250	$5,609,009
251–500	$18,241,667
>500	$19,980,556
Average for all respondents	$8,158,448

The average security budget in U.S. dollars, the average security budget as a percentage of organizational revenue, and the average security budget per organization employee, broken down by the size of the organization in terms of employee count. An average for all respondents is also included.

Organization Employee Count	Average Security Budget	Average Security Budget as a Percentage of Revenue	Average Security Budget Funds Per Organization Employee
<2,500	$2,617,262	0.045%	$1,930
2,500–5,000	$2,743,333	0.076%	$709
5,001–10,000	$4,060,524	0.095%	$532
10,001–25,000	$4,741,176	0.081%	$289
25,001–50,000	$8,176,801	0.039%	$243
>50,000	$25,103,516	0.079%	$198
Average for all respondents	$8,158,448	0.068%	$249

The average security budget in U.S. dollars, the average security budget as a percentage of organizational revenue, and the average security budget per organization employee, broken down by the size of the organization in terms of revenue. The average for all respondents is also included.			
Organization Revenue	**Average Security Budget**	**Average Security Budget as a Percentage of Revenue**	**Average Security Budget Funds Per Organization Employee**
<$500 million	$2,141,885	0.792%	$735
$500 million–$1 billion	$3,564,247	0.440%	$389
$1–$2.5 billion	$3,283,880	0.156%	$278
$2.5–$7.5 billion	$7,893,350	0.135%	$277
$7.5–$25 billion	$9,602,302	0.062%	$222
>$25 billion	$23,411,120	0.046%	$226
Average for all respondents	$8,158,448	0.068%	$249

The average security budget, broken down by the scope of the organization. The average for all respondents is also included.	
Organization Scope	**Average Security Budget**
International	$9,987,378
National	$6,113,711
Regional	$3,767,847
Local	$1,925,667
Average for all respondents	$8,158,448

For the purpose of breaking down the data by industry, a more detailed analysis than what appears in the following table was not appropriate for this survey due to a lack of sufficient respondents in all industries. Instead the table is derived from a selected list of industries where the numbers of respondents were adequate to report preliminary results. For the "government" industry category, the figure for budget as a percentage of revenue represents the average security budget as a percentage of the total appropriated budget for the government entity.

The average security budget in U.S. dollars, the average security budget as a percentage of organizational revenue, and the average security budget per organization employee, broken down by the organization's industry. The average for all respondents is also included.

Industry	Average Security Budget	Average Security Budget as a Percentage of Revenue	Average Security Budget Funds Per Organization Employee
Manufacturing	$3,509,501	0.060%	$195
Retail	$8,243,587	0.109%	$216
Banking, finance, insurance	$3,808,261	0.023%	$136
Energy, mining, utilities	$6,949,505	0.042%	$316
Government	$14,783,364	0.340%	$485
Average for all respondents	$8,158,448	0.068%	$249

The average security budget in U.S. dollars and the average security budget as a percentage of organizational revenue, broken down by the title level of the respondent. The average for all respondents is also included.

Title Level of Respondent	Average Security Budget	Average Security Budget as a Percentage of Revenue
Vice president or greater	$10,977,627	0.076%
Director	$10,435,321	0.086%
Senior manager	$5,188,475	0.053%
Manager or lower	$2,903,556	0.025%
Average for all respondents	$8,158,448	0.068%

A complete analysis by functional area was not valuable because some functional areas had too few respondents to consider the results reliable. However, in the following table, four selected functional areas that did have adequate respondents are analyzed.

The average security budget in U.S. dollars, the average security budget as a percentage of organizational revenue, and the average security budget per organization employee, broken down by the functional area in to which the respondent reports. The average for all respondents is also included.			
Functional Area in to Which Respondent Reports	**Average Security Budget**	**Average Security Budget as a Percentage of Revenue**	**Average Security Budget Funds Per Organization Employee**
Executive	$4,600,077	0.107%	$343
Facilities/real estate	$3,827,113	0.033%	$173
Financial	$7,270,000	0.072%	$168
Information technology	$5,760,000	0.023%	$121
Average for all respondents	$8,158,448	0.068%	$249

In the following table showing security budget by the respondent's scope of responsibility, the category "regional (consisting of multiple countries)" is excluded due to a low number of data points.

The average security budget and average security budget as a percentage of organizational revenue, broken down by the respondent's scope of responsibility. The average for all respondents is also included.		
Respondent's Scope of Responsibility	**Average Security Budget**	**Average Security Budget as a Percentage of Revenue**
All	$11,991,380	0.100%
International	$7,330,658	0.074%
National	$12,583,341	0.068%
Regional (within a country)	$3,143,565	0.026%
Local	$2,327,214	0.042%
Average for all respondents	$8,158,448	0.068%

SECTION IV: SECURITY EMPLOYEES

In this section, the security staff is examined. Unless otherwise noted, data included both outsourced and contracted security staff. Notable commonalities about security staff among all respondents include:

- On average, respondents reported one security employee for every 99 company employees.
- On average, a typical respondent managed a staff made up of 85 percent uniformed security.

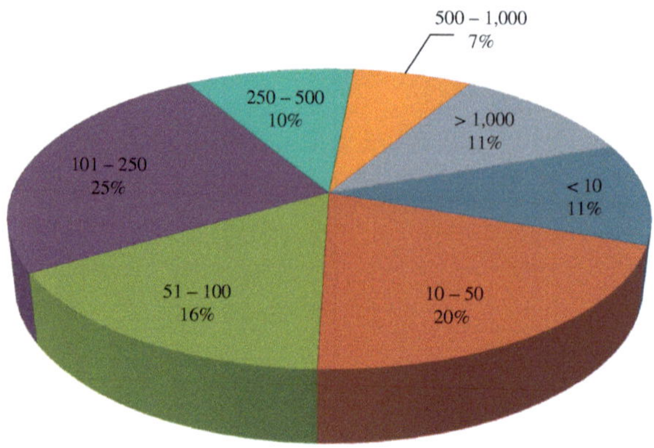

Figure 4.01 The number of security employees per organization (including any outsourced workforce).

A complete analysis by industry was not possible because some industries had too few respondents to consider the results reliable. However, in the following table, five industries that did have an adequate number of respondents are analyzed.

Industry	**The Average Number of Company Employees Per Security Employee**
Manufacturing	98
Retail	77
Banking, finance, insurance	120
Energy, mining, utilities	34
Government	130
Average for all respondents	99

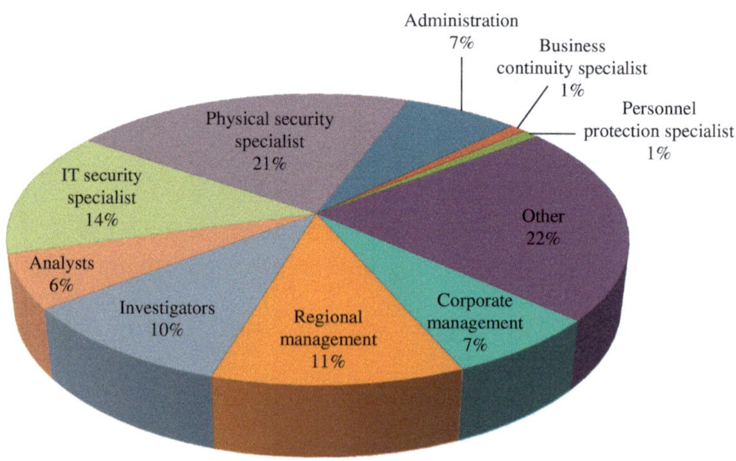

Figure 4.02 The types of security personnel managed by respondents. This data excludes any uniformed security staff.

The number of each type of security employee managed as a percentage (%) of the total security employees managed by each respondent. Data is categorized by the organization's overall employee count. For example, of the respondents who identified their organizations as having more than 50,000 total employees, 82 percent of their security staff would be considered uniformed security.

Organization Employee Count	Corp. Mgmt.	Regional Mgmt.	Investigators	Analysts	IT Security Spec.	Physical Security Spec.	Uniformed Security	Admin.	BC Spec.	Personnel Protection Spec.	Other
<2,500	0.8	1.4	1.1	0.4	0.7	4.5	86.6	1.7	0.3	0.1	2.5
2,500–5,000	1.3	4.5	1.3	1.2	4.4	2.3	82.7	1.5	0.0	0.2	0.6
5,001–10,000	1.7	1.5	1.3	1.4	2.7	11.0	78.6	1.3	0.2	0.2	0.1
10,001–25,000	1.2	0.6	2.3	0.5	0.9	3.6	88.8	1.3	0.1	0.1	0.7
25,001–50,000	1.5	2.4	1.1	1.1	6.1	0.8	86.3	0.4	0.2	0.1	0.0
>50,000	0.6	2.0	0.9	1.2	1.6	1.7	82.3	1.0	0.1	0.1	8.4
For all respondents	1.1	1.7	1.4	0.9	2.1	3.2	84.7	1.1	0.1	0.1	3.4

The number of each type of security employee managed as a percentage (%) of the total security employees managed by each respondent. Data is categorized by the organization's annual revenue.

Organization Revenue	Corp. Mgmt.	Regional Mgmt.	Investigators	Analysts	IT Security Spec.	Physical Security Spec.	Uniformed Security	Admin.	BC Spec.	Personnel Protection Spec.	Other
<$500M	1.5	1.8	0.8	0.3	1.3	3.0	84.8	1.5	0.3	0.2	4.6
$500M–$1B	0.9	1.9	0.5	0.4	1.6	3.1	91.0	0.3	0.1	0.1	0.0
$1B–$2.5B	1.7	0.8	1.0	1.4	3.1	4.4	85.0	1.4	0.1	0.2	0.9
$2.5B–$7.5B	2.6	1.5	6.2	1.6	2.2	9.1	67.1	3.1	0.1	0.5	6.0
$7.5B–$25B	1.1	2.1	1.5	0.9	4.0	1.9	87.0	0.8	0.1	0.1	0.4
>$25B	0.3	1.8	0.4	0.9	1.1	1.6	85.7	0.8	0.1	0.0	7.2
For all respondents	1.1	1.7	1.4	0.9	2.1	3.2	84.7	1.1	0.1	0.1	3.4

The number of each type of security employee managed as a percentage (%) of the total security employees managed by each respondent. Data is categorized by the functional area in to which the respondent reports.

Functional Area in to Which Respondent Reports	Corp. Mgmt.	Regional Mgmt.	Investigators	Analysts	IT Security Spec.	Physical Security Spec.	Uniformed Security	Admin.	BC Spec.	Personnel Protection Spec.	Other
Administration	3.2	1.5	1.0	0.5	0.5	0.7	89.3	3.4	0.0	0.0	0.0
Environmental, health, and safety	0.3	1.8	0.1	0.0	1.1	2.1	93.9	0.6	0.0	0.1	0.0
Executive	1.2	2.2	2.7	0.8	3.3	6.4	81.8	1.2	0.0	0.1	0.3
Facilities/real estate	3.5	1.3	0.7	1.0	1.4	4.4	85.0	2.2	0.3	0.2	0.1
Finance	1.1	2.8	1.6	0.5	1.5	0.8	89.6	0.7	0.0	0.1	1.2
Human resources	1.6	0.7	0.7	0.0	0.4	2.9	91.3	1.8	0.0	0.5	0.0
Information technology	2.3	4.7	3.7	12.9	13.7	8.9	51.6	2.3	0.0	0.0	0.0
Legal	0.5	0.4	1.0	0.8	0.8	1.7	93.8	0.5	0.1	0.1	0.2
Operations	0.3	0.3	0.3	0.1	0.9	0.0	98.0	0.1	0.0	0.0	0.0
Risk, audit, compliance	2.6	1.9	3.2	0.6	1.6	3.9	82.9	2.9	0.3	0.0	0.0
Shared services	0.5	0.5	0.4	0.4	2.5	2.7	67.1	1.6	0.7	0.0	23.6
Other	0.6	3.2	0.7	0.8	0.5	1.1	85.8	1.0	0.0	0.3	6.1
For all respondents	1.1	1.7	1.4	0.9	2.1	3.2	84.7	1.1	0.1	0.1	3.4

SECTION V: BOARD-LEVEL RISK CATEGORY BREAKDOWN

The Security Executive Council's board-level risk research examines typical concerns by boards with security issues and matches them to typical risk mitigation strategies applied by the security function. As discussed in the March 2010 issue of *Corporate Secretary*, "... the failure to recognize the potential consequences of risk failure can bring serious damage to companies."[1] Security and risk management leaders must be able to effectively communicate security's role in mitigating board-level risk concerns to the board and executive management team.

Survey respondents were allowed to designate up to five risks that were of greatest concern to their organization. For the purpose of this next section of analysis, verbatim responses were categorized based on their best-fitting board-level risk category (see Appendix A for a full description of board-level risk categories). The charts below can be used to identify the board-level risk categories that were most commonly chosen as high risks by the respondents.

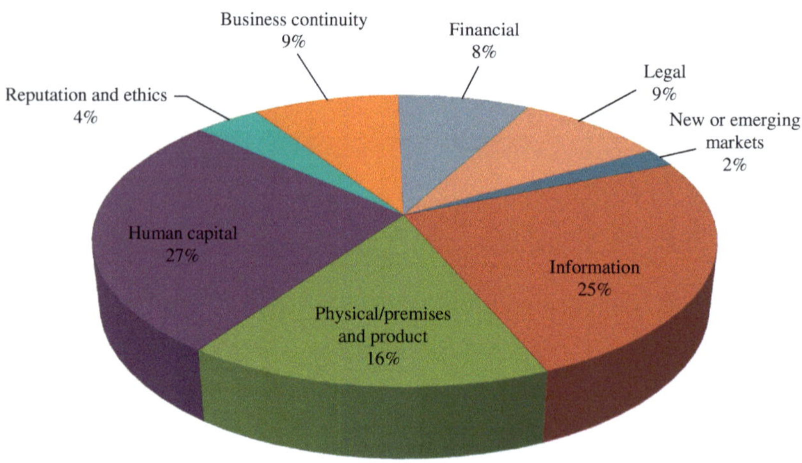

Figure 5.01 The most commonly identified board-level risks.

Analysis of board-level risk category showed no correlation to the size of the organization when broken down by employee count or revenue. Because, little value could be derived from organization size analysis, those charts are excluded from this report.

Other factors, such as organization scope, industry type, the respondent's title level, the respondent's level of reporting away from the senior-most operating executive, and the functional area in to which the respondent reports did reveal certain commonalities of board-level risk concern. These analyses are included in the tables below.

The occurrence (%) of the specified categories of board-level risk by organization scope. For example, for respondents in organizations with international scope, the top risk concern was information, at 27 Percent.

Organization Scope	Information	Physical/ Premises and Product	Human Capital	Reputation and Ethics	Business Continuity	Financial	Legal	New or Emerging Markets
International	27	14	26	5	10	7	9	3
National	15	26	28	2	11	11	6	0
Regional (within a nation)	37	7	22	4	4	19	7	0
Local	14	28	33	3	3	6	14	0
For all respondents	25	16	26	4	9	8	9	2

The occurrence (%) of the specified categories of board-level risk by industry.

Industry	Information	Physical/ Premises and Product	Human Capital	Reputation and Ethics	Business Continuity	Financial	Legal	New or Emerging Markets
Business services	50	0	23	0	14	0	14	0
Manufacturing	30	13	30	4	7	9	5	2
Retail	13	23	17	3	7	17	10	10
Banking, finance, insurance	32	7	28	6	8	13	6	1
Energy, mining, utilities	17	17	26	3	9	14	9	6
Pharmaceuticals	35	18	24	6	6	0	12	0
Real estate	0	25	17	8	33	0	17	0
Government	26	22	26	2	11	2	11	0
Transportation services	35	15	20	15	0	5	5	5
Health care	23	23	27	0	9	5	14	0
Food, beverage	7	33	33	11	0	11	0	4
For all respondents	25	16	26	4	9	8	9	2

It should be kept in mind that there were not enough respondents in this survey to make definitive statements regarding risk by industry. However, in the table illustrating the occurrence of board-level risk categories by industry, it appears that most concerns fell in the human capital and information board-level risk categories, followed by physical/premises and product.

The occurrence (%) of the specified categories of board-level risk by title level of respondent.

Title Level of Respondent	Information	Physical/ Premises and Product	Human Capital	Reputation and Ethics	Business Continuity	Financial	Legal	New or Emerging Markets
Vice president, executive vice president, senior vice president, chief security officer	28	11	26	3	9	12	9	1
Director	30	12	28	5	9	4	8	4
Senior manager	22	22	29	4	4	10	7	1
Manager	9	25	20	5	13	13	15	0
For all respondents	25	16	26	4	9	8	9	2

The occurrence (%) of the specified categories of board-level risk by the levels of reporting between the respondent and the senior-most operating executive.

Levels of Reporting away from Senior-Most Operating Executive (CEO or Equivalent)	Information	Physical/ Premises and Product	Human Capital	Reputation and Ethics	Business Continuity	Financial	Legal	New or Emerging Markets
1 (direct reporting relationship)	21	11	33	6	9	9	3	8
2 levels	23	16	25	5	11	10	9	1
3 levels	31	23	23	2	4	4	9	2
More than 3 levels	27	15	27	3	8	6	13	0
For all respondents	25	16	26	4	9	8	9	2

The occurrence (%) of the specified categories of board-level risk by the functional area in to which the respondent reports.								
Functional Area in to Which Respondent Reports	**Information**	**Physical/ Premises and Product**	**Human Capital**	**Reputation and Ethics**	**Business Continuity**	**Financial**	**Legal**	**New or Emerging Markets**
Admin.	10	20	20	10	20	20	0	0
Environmental, health, and safety	7	47	20	0	7	7	13	0
Executive	29	16	23	5	10	8	7	2
Facilities/real estate	14	19	46	2	12	3	5	0
Finance	10	19	38	0	10	5	5	14
Human resources	13	25	25	0	0	13	25	0
Information technology	61	5	9	2	2	0	20	0
Legal	18	13	25	8	8	15	10	5
Operations	20	20	33	13	0	0	7	7
Risk, audit, compliance	21	0	29	0	29	14	7	0
Shared services	19	15	30	7	7	15	7	0
Other	28	22	19	6	6	11	6	3
For all respondents	25	16	26	4	9	8	9	2

SECTION VI: PROGRAM DRIVERS

Organizations have different key motivators for their risk management programs. Even within a particular organization different security programs may have different drivers, depending on what risks they are addressing. While one type of driver is not necessarily better than another, it is interesting to note which drivers are more prevalent than others.

Respondents were allowed to select multiple program drivers. Consequently, the numbers shown in Figure 6.01 indicate the program

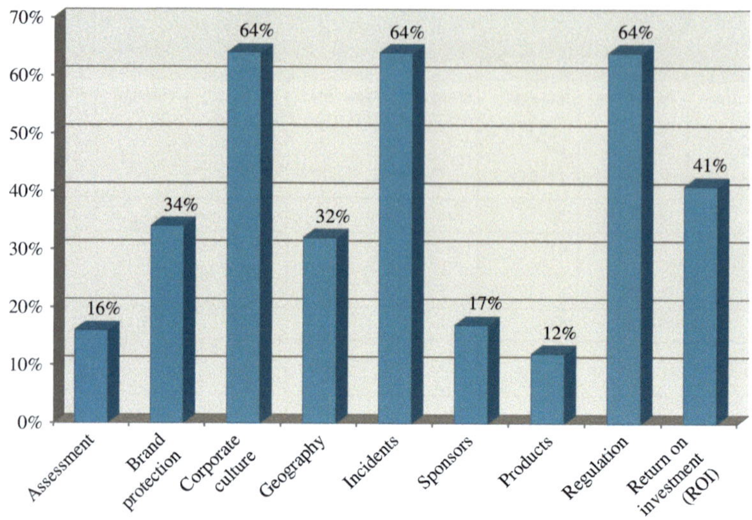

Figure 6.01 The frequency (%) that a given security program driver was identified by respondents as a key motivator.

drivers most often selected by the respondents, which is not necessarily the most *important* program driver to the respondents. The most common drivers for security programs were regulations, corporate culture, and incidents.

Results are further broken down by organization scope and size, industry type, the respondent's title level, the respondent's level of reporting away from the senior-most operating executive, and the functional area in to which the respondent reports. These analyses are included in the following tables.

The frequency (%) that a given security program driver was identified by respondents as a key motivator, broken down by organization scope. For example, of respondents whose organizations have a national scope, 76 percent identified corporate culture as security program driver.

Organization Scope	Assessment	Brand Protection	Corporate Culture	Geography	Incidents	Sponsors	Products	Regulation	Return on Investment (ROI)
International	13	39	60	42	60	14	14	64	44
National	19	29	76	14	62	24	10	57	43
Regional (within a nation)	29	14	86	0	100	14	0	57	14
Local	20	20	60	10	80	30	10	80	30
For all respondents	16	34	64	32	64	17	12	64	41

The frequency (%) that a given security program driver was identified by respondents as a key motivator,broken down by organization size in terms of employee count. (Note: The data for all respondents in this table may not match the data in Figure 6.01 because organizations without known employee counts are excluded from this table.)

Organization Employee Count	Assessment	Brand Protection	Corporate Culture	Geography	Incidents	Sponsors	Products	Regulation	Return on Investment (ROI)
<2,500	19	44	81	19	63	19	6	38	31
2,500–5,000	18	35	76	35	76	18	6	65	65
5,001–10,000	28	28	60	20	68	20	4	76	32
10,001–25,000	13	29	50	38	67	29	17	75	33
25,001–50,000	6	56	61	33	72	17	17	50	28
>50,000	9	23	64	45	41	0	23	68	59
For all respondents	16	34	64	32	64	17	12	64	41

The frequency (%) that a given security program driver was identified by respondents as a key motivator, broken down by organization size in terms of annual revenue. (Note: The data for all respondents in this table may not match the data in Figure 6.01 because organizations without known revenue are excluded from this table.)

Organization Revenue	Assessment	Brand Protection	Corporate Culture	Geography	Incidents	Sponsors	Products	Regulation	Return on Investment (ROI)
<$500M	25	33	75	17	75	25	8	58	33
$500M–$1B	13	27	47	33	60	20	0	73	33
$1B–$2.5B	19	26	78	19	70	15	4	63	48
$2.5B–$7.5B	21	42	50	42	63	17	17	58	33
$7.5B–$25B	12	42	69	31	73	19	15	69	42
>$25B	6	38	69	56	44	13	31	56	56
For all respondents	16	35	65	33	65	18	13	63	42

The frequency (%) that a given security program driver was identified by respondents as a key motivator, broken down by industry. (Note: Industries with a small number of respondents are excluded from this table.)

Industry	Assessment	Brand Protection	Corporate Culture	Geography	Incidents	Sponsors	Products	Regulation	Return on Investment (ROI)
Business services	14	14	43	43	29	14	14	71	0
Manufacturing	6	47	76	35	76	6	29	41	41
Retail	11	33	56	11	56	22	11	67	44
Banking, finance, insurance	21	38	67	17	75	25	4	63	33
Energy, mining, utilities	22	56	56	67	67	11	11	89	56
Pharmaceuticals	0	33	67	33	67	17	17	83	33
Government	21	21	57	21	71	21	7	93	29
Health care	17	17	50	0	50	33	0	67	17
Food, beverage	33	67	83	83	83	33	33	0	83
For all respondents	16	34	64	32	64	17	12	64	41

The frequency (%) that a given security program driver was identified by respondents as a key motivator, broken down by the title level of the respondent.

Title Level of Respondent	Assessment	Brand Protection	Corporate Culture	Geography	Incidents	Sponsors	Products	Regulation	Return on Investment (ROI)
Vice president, executive vice president, senior vice president, chief security officer	8	33	63	33	50	13	13	63	54
Director	14	43	53	37	67	18	10	73	35
Senior manager	24	28	80	28	68	20	20	44	40
Manager	11	21	79	21	63	11	11	63	47
For all respondents	16	34	64	32	64	17	12	64	41

The frequency (%) that a given security program driver was identified by respondents as a key motivator, broken down by the levels of reporting between the respondent and the senior-most operating executive.

Levels of Reporting away from Senior-Most Operating Executive (CEO or Equivalent)	Assessment	Brand Protection	Corporate Culture	Geography	Incidents	Sponsors	Products	Regulation	Return on Investment (ROI)
1 (direct reporting relationship)	21	47	63	47	63	11	26	63	32
2 levels	13	25	67	35	58	18	9	64	40
3 levels	14	39	68	29	71	18	7	57	54
More than 3 levels	20	40	50	15	70	20	15	75	35
For all respondents	16	34	64	32	64	17	12	64	41

The frequency (%) that a given security program driver was identified by respondents as a key motivator, broken down by the functional area in to which the respondent reports.

Functional Area in to Which Respondent Reports	Assessment	Brand Protection	Corporate Culture	Geography	Incidents	Sponsors	Products	Regulation	Return on Investment (ROI)
Admin.	33	67	67	0	100	0	0	67	67
Environmental, health, and safety	0	0	75	25	100	0	25	50	50
Executive	14	31	66	31	69	14	11	69	17
Facilities/real estate	24	41	76	35	71	24	24	65	65
Finance	0	50	63	50	50	13	0	38	50
Human resources	0	25	100	25	50	25	0	25	50
Information technology	7	29	43	14	64	14	14	79	43
Legal	27	36	64	36	64	36	27	64	36
Operations	0	0	67	67	67	0	0	67	100
Risk, audit, compliance	20	20	60	20	40	0	0	40	40
Shared services	25	50	50	50	38	25	0	88	38
Other	20	40	60	30	60	20	10	60	50
For all respondents	16	34	64	32	64	17	12	64	41

SECTION VII: SECURITY PROGRAM GOVERNANCE

In this section, the various methods by which security programs are governed are examined. **Respondents were allowed to select multiple methods**. The data illustrated in Figure 7.01 indicate the governance method most often selected by the respondents, which is not necessarily the most important method to the respondents.

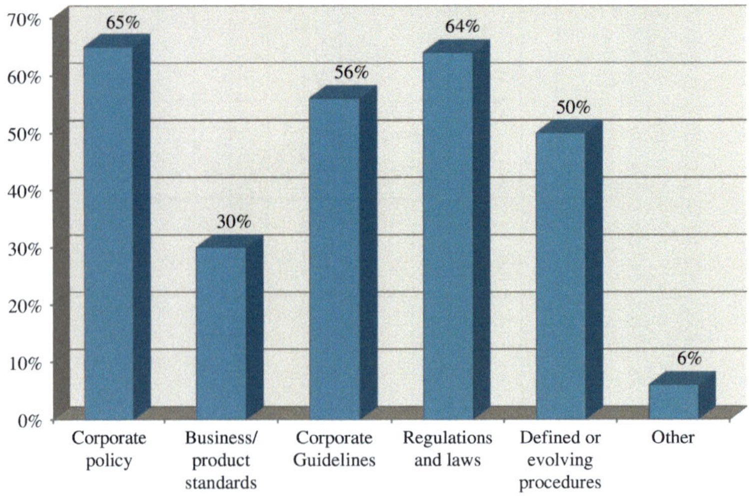

Figure 7.01 The percentage of respondents that selected a specific method of governance for the security program.

Governance methods provided as "other" included:

- Comparable company practices
- Bylaws and board of directors
- Client mandates

Results are further broken down by organization scope and size, industry type, the respondent's title level, the respondent's level of reporting away from the senior-most operating executive, and the functional area in to which the respondent reports. These analyses, which revealed certain commonalities of board-level risk concern, are included in the following tables.

The percentage (%) of respondents that selected a specific method of governance for the security program, broken down by organization scope. For example, of respondents that belong to a regional organization, 29 percent stated their security program was at least partially governed by corporate guidelines.

Organization Scope	Corporate Policy	Business/Product Standards	Corporate Guidelines	Regulations and Laws	Defined or Evolving Procedures
International	63	32	56	64	49
National	57	29	67	62	67
Regional (within a nation)	100	14	29	57	29
Local	70	30	50	70	40
For all respondents	65	30	56	64	50

The percentage (%) of respondents that selected a specific method of governance for the security program, broken down by organization size in terms of employee count. (Note: The totals in this table may not match the data in Figure 7.01 because organizations without known employee counts are excluded from this table.)

Organization Employee Count	Corporate Policy	Business/ Product Standards	Corporate Guidelines	Regulations and Laws	Defined or Evolving Procedures
<2,500	63	31	56	69	50
2,500–5,000	53	41	53	59	76
5,001–10,000	80	20	44	64	32
10,001–25,000	58	17	50	75	33
25,001–50,000	61	33	61	56	56
>50,000	68	45	73	59	64
For all respondents	65	30	56	64	50

The percentage (%) of respondents that selected a specific method of governance for the security program, broken down by organization size in terms of revenue. (Note: The totals in this table may not match the data in Figure 7.01 because organizations without known revenue are excluded from this table.)

Organization Revenue	Corporate Policy	Business/Product Standards	Corporate Guidelines	Regulations and Laws	Defined or Evolving Procedures
<$500M	83	33	58	75	42
$500M–$1B	53	27	27	67	67
$1B–$2.5B	70	37	67	52	44
$2.5B–$7.5B	63	17	42	71	38
$7.5B–$25B	62	19	62	65	54
>$25B	63	63	81	56	69
For all respondents	65	31	57	63	51

The percentage (%) of respondents that selected a specific method of governance for the security program, broken down by industry. (Note: Industries with a small number of respondents are excluded from this table.)

Industry	Corporate Policy	Business/Product Standards	Corporate Guidelines	Regulations and Laws	Defined or Evolving Procedures
Business services	57	57	43	71	43
Manufacturing	65	29	65	41	59
Retail	56	22	56	56	33
Banking, finance, insurance	83	33	50	71	33
Energy, mining, utilities	56	0	56	78	33
Government	64	14	36	86	36
Health care	67	17	33	67	50
Food, beverage	67	67	83	50	83
For all respondents	65	31	57	63	51

The percentage (%) of respondents that selected a specific method of governance for the security program, broken down by the title level of the respondent.					
Title Level of Respondent	**Corporate Policy**	**Business/ Product Standards**	**Corporate Guidelines**	**Regulations and Laws**	**Defined or Evolving Procedures**
Vice president, executive vice president, senior vice president, chief security officer	67	33	54	54	54
Director	63	18	53	73	45
Senior manager	72	28	60	48	56
Manager	58	58	63	63	58
For all respondents	65	30	56	64	50

The percentage (%) of respondents that selected a specific method of governance for the security program, broken down by the number of levels between the respondent and the senior-most operating executive.					
Levels of Reporting away from Senior Most Operating Executive (CEO or Equivalent)	**Corporate Policy**	**Business/ Product Standards**	**Corporate Guidelines**	**Regulations and Laws**	**Defined or Evolving Procedures**
1 (direct reporting relationship)	58	5	58	58	53
2 levels	65	31	58	67	53
3 levels	64	43	54	54	61
More than 3 levels	70	35	50	75	25
For all respondents	65	30	56	64	50

The percentage (%) of respondents that selected a specific method of governance for the security program, broken down by the functional area in to which the respondent reports.					
Functional Area in to Which Respondent Reports	**Corporate Policy**	**Business/ Product Standards**	**Corporate Guidelines**	**Regulations and Laws**	**Defined or Evolving Procedures**
Admin.	100	0	33	100	0
Environmental, health, and safety	100	50	100	100	75
Executive	60	20	54	69	57
Facilities/real estate	53	41	65	59	59
Finance	63	13	50	38	38
Human resources	50	0	100	75	50
Information technology	79	57	43	64	36
Legal	73	18	55	64	36
Operations	33	33	67	67	33
Risk, audit, compliance	60	20	40	40	40
Shared services	75	63	50	50	50
Other	60	30	50	70	70
For all respondents	65	30	56	64	50

SECTION VIII: RISK OVERSIGHT

In this section, the different types of corporate risk oversight groups that existed within the respondents' organizations are identified and analyzed (see Figure 8.01).

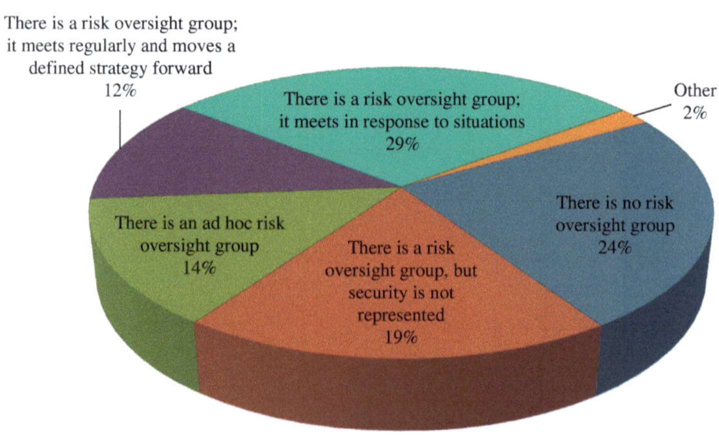

Figure 8.01 The types of risk oversight groups that currently exist in the respondents' organizations.

The percentage (%) of a given type of risk oversight group in the respondents' organizations, broken down by organization size in terms of employee count. (Note: The totals in this table may not match the totals in Figure 8.01 because organizations without employee counts are excluded from this table.)						
Organization Employee Count	**There Is No Risk Oversight Group**	**There Is a Risk Oversight Group, but Security Is Not Represented**	**There Is an Ad Hoc Risk Oversight Group**	**There Is a Risk Oversight Group; It Meets in Response to Incidents**	**There Is a Risk Oversight Group; It Meets Regularly and Moves a Defined Strategy Forward**	**Other**
<2,500	31	13	13	25	19	0
2,500–5,000	41	12	6	6	35	0
5,001–10,000	20	24	12	4	36	4
10,001–25,000	25	13	25	17	21	0
25,001–50,000	11	22	17	11	33	6
>50,000	18	27	9	14	27	5
For all respondents	24	19	14	12	29	2

The percentage (%) of a given type of risk oversight group in the respondents' organizations, broken down by organization size in terms of annual revenue. (Note: The totals in this table may not match the totals in Figure 8.01 because organizations without known revenue are excluded from this table.)						
Organization Revenue	**There Is No Risk Oversight Group**	**There Is a Risk Oversight Group, but Security Is Not Represented**	**There Is an Ad Hoc Risk Oversight Group**	**There Is a Risk Oversight Group; It Meets in Response to Incidents**	**There Is a Risk Oversight Group; It Meets Regularly and Moves a Defined Strategy Forward**	**Other**
<$500M	8	8	8	33	42	0
$500M–$1B	40	7	20	7	20	7
$1B–$2.5B	33	19	15	4	26	4
$2.5B–$7.5B	33	25	17	8	13	4
$7.5B–$25B	4	31	12	23	31	0
>$25B	25	13	6	6	50	0
For all respondents	24	19	13	13	28	3

It may be expected that smaller organizations or those with smaller revenues would tend not to have well-defined risk oversight groups, but the data displayed in the following two tables could neither confirm nor refute that expectation.

If a risk oversight group did exist in his or her organization, the respondent was asked which business functions were represented in the group. **Respondents could choose more than one business function.** The data in Figure 8.02 indicates the business functions most often selected by the respondents, which are not necessarily the most important method to the respondents. No clear correlation emerged between the size of the organization and the business functions that were included in the risk oversight groups.

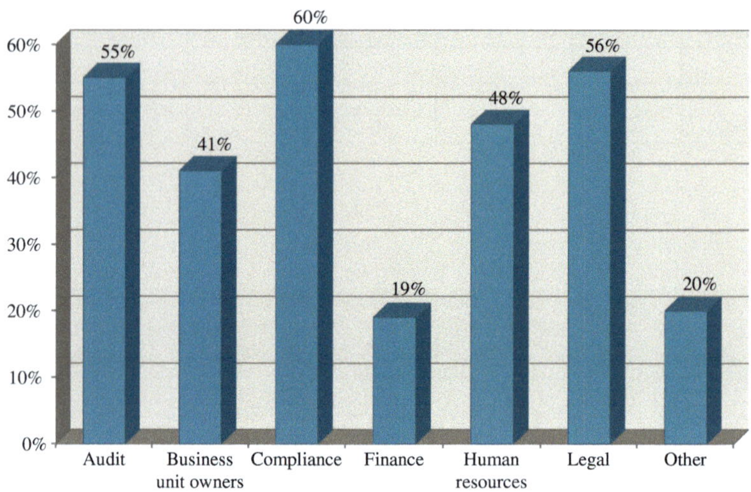

Figure 8.02 The frequency (%) with which specific business functions were involved in the risk oversight group.

Business functions provided as "other" included:

- Business continuity management
- Insurance
- Board of directors
- Chief information officer/information technology

In the following tables, the results in Figure 8.02 are further broken down by organization size, industry type, and the functional area in to which the respondent reports.

The percentage (%) of risk oversight groups that involve specific business functions, broken down by organization size in terms of employee count. (Note: The totals in this table may not match the totals in Figure 8.02 because organizations without employee counts are excluded from this table.)						
Organization Employee Count	**Audit**	**Business Unit Owners**	**Compliance**	**Finance**	**Human Resources**	**Legal**
<2,500	38	31	50	6	50	38
2,500–5,000	47	24	59	12	41	53
5,001–10,000	52	48	60	36	48	56
10,001–25,000	58	42	63	17	50	67
25,001–50,000	61	44	67	28	50	67
>50,000	68	50	59	9	45	50
For all respondents	55	41	60	19	48	56

The percentage (%) of risk oversight groups that involve specific business functions, broken down by organization size in terms of annual revenue. (Note: The totals in this table may not match the totals in Figure 8.02 because organizations without known revenue are excluded from this table.)						
Organization Revenue	**Audit**	**Business Unit Owners**	**Compliance**	**Finance**	**Human Resources**	**Legal**
<$500M	67	58	83	17	75	75
$500M–$1B	47	40	47	7	40	47
$1B–$2.5B	44	37	48	22	48	52
$2.5B–$7.5B	50	29	58	21	42	46
$7.5B–$25B	65	42	69	27	54	69
>$25B	63	50	56	13	38	50
For all respondents	55	41	59	19	48	56

The percentage (%) of risk oversight groups that involve specific business functions, broken down by industry.

Industry	Audit	Business Unit Owners	Compliance	Finance	Human Resources	Legal
Business services	57	43	57	14	43	71
Manufacturing	47	35	53	6	53	65
Retail	67	22	44	0	44	44
Banking, finance, insurance	67	67	88	13	50	63
Energy, mining, utilities	67	56	67	44	56	56
Government	36	36	43	21	36	36
Health care	67	33	50	0	50	50
Food, beverage	50	33	67	67	67	67
For all respondents	55	41	60	19	48	56

The percentage (%) of risk oversight groups that involve specific business functions, broken down by the functional area in to which the respondent reports.

Functional Area in to Which Respondent Reports	Audit	Business Unit Owners	Compliance	Finance	Human Resources	Legal
Admin.	33	67	67	0	33	33
Environmental, health, and safety	50	50	75	0	50	75
Executive	46	43	57	23	54	57
Facilities/real estate	47	24	41	12	41	53
Finance	88	63	88	13	38	50
Human resources	75	25	75	25	50	75
Information technology	71	50	64	7	36	50
Legal	55	55	45	45	55	64
Operations	67	0	33	0	0	33
Risk, audit, compliance	40	40	100	20	20	40
Shared services	50	25	63	0	63	63
Other	60	40	60	40	70	60
For all respondents	55	41	60	19	48	56

SECTION IX: SECURITY PROGRAMS

In this section, the level of responsibility respondents stated they had for selected programs or services at their organization is analyzed. For a given program, respondents were asked to describe their level of responsibility in the following terms:

- Accountable
- Responsible
- Accountable and responsible
- Outside scope (meaning the program or service existed within the organization but the respondent was neither accountable nor responsible for it)
- Program does not exist

Any respondents who did not provide answers to this portion of the survey were removed from this section's analysis.

The following table shows the percentage of respondents who selected a particular level of responsibility for the given program or service. The most common answer for the given program or service is highlighted.

The percentage (%) of all respondents who were accountable, responsible, accountable and responsible, or neither accountable nor responsible for a given security program or service. Respondents could also choose "program does not exist."

Program	Accountable	Responsible	Accountable and Responsible	Outside Scope	Program Does Not Exist
Asset protection, loss prevention	10	14	64	8	4
Aviation security and operations	4	7	10	33	45
Awareness and education	13	24	57	2	4
Brand, reputation, trademark/ trade name	13	23	10	39	15
Business continuity/resiliency	10	20	31	32	6
Business ethics and compliance	10	28	5	47	10
Business intelligence and analysis	6	14	10	56	14

(Continued)					
Program	**Accountable**	**Responsible**	**Accountable and Responsible**	**Outside Scope**	**Program Does Not Exist**
Emergency response and disaster recovery	10	22	46	19	2
Guard service	2	2	11	4	81
Incident reporting	6	18	70	3	2
Information security, application security	6	13	18	53	10
Information security, intellectual property protection	9	15	22	44	10
Information security, intrusion detection	6	10	17	56	10
Information security, systems and specifications: hardware	2	6	17	61	14
Information security, systems and specifications: software	3	6	16	61	14
Information technology forensic investigation	8	8	19	54	11
Investigations (e.g. theft, fraud, threats, due diligence, conflict of interest)	10	17	57	10	5
Law enforcement liaison	5	7	71	12	5
New product and marketing security	9	11	16	40	24
Personnel protection/workplace violence	11	10	56	17	6
Personnel screening, background investigations, and due diligence	9	18	23	42	9
Physical site security design and system specifications	9	7	66	13	5
Product protection (e.g. tampering, counterfeiting, diversion)	8	7	15	42	28
R&D security	13	7	27	23	31
Risk assessments	10	17	62	6	5
Security-related regulations and compliance management	6	11	77	2	3
Security/business command center	6	10	55	12	16
Supply chain security	8	14	23	31	25
Travel security	8	13	40	23	16
Others	1	2	6	6	85

In the following tables, the "most common" column indicates the most common level of responsibility and the percentage of respondents who claimed that level of responsibility. The other columns break down the responses into categories (including organization size, scope, and functional reporting area), and provide the most common response for that particular category along with the percentage of respondents who chose the response.

For example, in the following table, the most common response for an aviation security and operations program (45 percent of all respondents) was that the organization did not have one. For respondents reporting organization sizes of 25,000 to 50,000 employees, however, the most common response (41 percent of respondents) was that an aviation security and operations program exists in the organization, but the respondent is neither accountable nor responsible for it.

Table Key	
ACCT	Respondent is accountable for the program.
RESP	Respondent is responsible for the program.
BOTH	Respondent is both accountable and responsible for the program.
SCOPE	Program exists in the organization, but the respondent is neither accountable nor responsible for it.
NONE	Program does not exist in the organization.

Use this table key to interpret the color coding used in the following four tables.

The most common level of responsibility indicated by respondents for a specific security program, categorized by organization size in Terms of employee count. The percentage (%) of respondents in each category who chose that level of responsibility is included in parentheses.

		Organization Size (in Terms of Employee Count)					
Program	**Most Common**	**>2,500**	**2,500–5,000**	**5,001–10,000**	**10,001–25,000**	**25,001–50,000**	**<50,000**
Asset protection, loss prevention	BOTH (64)	BOTH (81)	BOTH (76)	BOTH (48)	BOTH (52)	BOTH (65)	BOTH (67)
Aviation security and operations	NONE (45)	NONE (69)	NONE (53)	NONE (40)	NONE (39)	SCOPE (41)	SCOPE (43)
Awareness and education	BOTH (58)	BOTH (63)	BOTH (71)	BOTH (60)	BOTH (39)	BOTH (53)	BOTH (57)
Brand, reputation, trademark/trade name	SCOPE (39)	SCOPE (38)	SCOPE (53)	RESP (36)	SCOPE (39)	SCOPE (41)	SCOPE (48)
Business continuity/resiliency	SCOPE (33)	BOTH (50)	BOTH (41)	SCOPE (40)	RESP (39)	RESP (41)	SCOPE (43)
Business ethics and compliance	SCOPE (47)	SCOPE (38)	SCOPE (53)	SCOPE (40)	RESP (43)	SCOPE (53)	SCOPE (62)
Business intelligence and analysis	SCOPE (56)	SCOPE (50)	SCOPE (65)	SCOPE (52)	SCOPE (65)	SCOPE (59)	SCOPE (52)
Emergency response and disaster recovery	BOTH (46)	BOTH (63)	BOTH (47)	BOTH (48)	BOTH (39)	BOTH (35)	BOTH (43)
Guard service	NONE (81)	NONE (75)	NONE (76)	NONE (72)	NONE (87)	NONE (82)	NONE (90)
Incident reporting	BOTH (71)	BOTH (81)	BOTH (88)	BOTH (68)	BOTH (70)	BOTH (47)	BOTH (71)
Information security, application security	SCOPE (54)	SCOPE (56)	SCOPE (59)	SCOPE (56)	SCOPE (48)	SCOPE (47)	SCOPE (62)
Information security, intellectual property protection	SCOPE (45)	SCOPE (50)	SCOPE (59)	SCOPE (44)	SCOPE (43)	RESP (35)	SCOPE (43)
Information security, intrusion detection	SCOPE (57)	SCOPE (50)	SCOPE (65)	SCOPE (52)	SCOPE (65)	SCOPE (53)	SCOPE (57)
Information security, systems, and specifications: hardware	SCOPE (62)	SCOPE (63)	SCOPE (53)	SCOPE (60)	SCOPE (78)	SCOPE (59)	SCOPE (57)
Information security, systems, and specifications: software	SCOPE (62)	SCOPE (63)	SCOPE (59)	SCOPE (52)	SCOPE (78)	SCOPE (59)	SCOPE (62)
Information technology forensic investigation	SCOPE (54)	SCOPE (63)	SCOPE (65)	SCOPE (56)	SCOPE (52)	SCOPE (41)	SCOPE (52)
Investigations (e.g. theft, fraud, threats, due diligence, conflict of interest)	BOTH (58)	BOTH (81)	RESP (41)	BOTH (64)	BOTH (57)	BOTH (47)	BOTH (57)

Law enforcement liaison	BOTH (72)	BOTH (88)	BOTH (76)	BOTH (60)	BOTH (78)	BOTH (59)	BOTH (67)
New product and marketing security	SCOPE (40)	NONE (31)	SCOPE (59)	SCOPE (32)	SCOPE (39)	SCOPE (53)	SCOPE (38)
Personnel protection/workplace violence	BOTH (56)	BOTH (69)	BOTH (59)	BOTH (40)	BOTH (57)	BOTH (53)	BOTH (62)
Personnel screening, background investigations, and due diligence	SCOPE (42)	SCOPE (31)	SCOPE (59)	SCOPE (48)	SCOPE (30)	SCOPE (53)	SCOPE (38)
Physical site security design and system specifications	BOTH (67)	BOTH (94)	BOTH (82)	BOTH (48)	BOTH (70)	BOTH (59)	BOTH (62)
Product protection (e.g. tampering, counterfeiting, diversion)	SCOPE (42)	ACCT (31)	SCOPE (76)	SCOPE (36)	NONE (39)	SCOPE (71)	SCOPE (29)
R&D security	NONE (30)	BOTH (38)	BOTH (41)	NONE (44)	NONE (43)	SCOPE (35)	BOTH (29)
Risk assessments	BOTH (63)	BOTH (69)	BOTH (71)	BOTH (64)	BOTH (65)	RESP (35)	BOTH (71)
Security-related regulations and compliance management	BOTH (77)	BOTH (100)	BOTH (88)	BOTH (76)	BOTH (70)	BOTH (65)	BOTH (71)
Security/business command center	BOTH (55)	BOTH (56)	BOTH (53)	BOTH (64)	BOTH (48)	BOTH (65)	BOTH (48)
Supply chain security	SCOPE (31)	BOTH (31)	SCOPE (59)	NONE (40)	NONE (30)	SCOPE (47)	BOTH (33)
Travel security	BOTH (40)	BOTH (25)	SCOPE (41)	BOTH (36)	BOTH (39)	BOTH (53)	BOTH (62)
Others	NONE (85)	NONE (94)	NONE (71)	NONE (76)	NONE (96)	NONE (88)	NONE (81)

The most common level of responsibility indicated by respondents for a specific security program, categorized by organization size in terms of annual revenue. The percentage (%) of respondents in each category who chose that level of responsibility is included in parentheses.

	Organization Size (in Terms of Revenue)						
Program	**Most Common**	**< $500M**	**$500M–$1B**	**$1B–$2.5B**	**$2.5B–$7.5B**	**$7.5B–$25B**	**> $25B**
Asset protection, loss prevention	BOTH (64)	BOTH (64)	BOTH (60)	BOTH (74)	BOTH (70)	BOTH (46)	BOTH (67)
Aviation security and operations	NONE (45)	NONE (73)	NONE (60)	NONE (59)	NONE (35)	SCOPE (50)	SCOPE (47)
Awareness and education	BOTH (58)	BOTH (55)	BOTH (47)	BOTH (52)	BOTH (65)	BOTH (54)	BOTH (60)
Brand, reputation, trademark/trade name	SCOPE (39)	RESP (36)	SCOPE (40)	SCOPE (48)	SCOPE (26)	SCOPE (46)	SCOPE (40)
Business continuity/ resiliency	SCOPE (33)	BOTH (45)	SCOPE (33)	SCOPE (41)	SCOPE (35)	RESP (31)	BOTH (33)
Business ethics and compliance	SCOPE (47)	RESP (45)	SCOPE (60)	SCOPE (41)	SCOPE (43)	SCOPE (58)	SCOPE (60)
Business intelligence and analysis	SCOPE (56)	SCOPE (55)	SCOPE (53)	SCOPE (59)	SCOPE (57)	SCOPE (69)	SCOPE (47)
Emergency response and disaster recovery	BOTH (46)	BOTH (55)	BOTH (47)	BOTH (56)	BOTH (35)	BOTH (38)	BOTH (47)
Guard service	NONE (81)	NONE (64)	NONE (80)	NONE (81)	NONE (83)	NONE (77)	NONE (93)
Incident reporting	BOTH (71)	BOTH (64)	BOTH (73)	BOTH (81)	BOTH (65)	BOTH (73)	BOTH (60)
Information security, application security	SCOPE (54)	SCOPE (45)	SCOPE (53)	SCOPE (56)	SCOPE (48)	SCOPE (62)	SCOPE (60)
Information security, intellectual property protection	SCOPE (45)	SCOPE (55)	SCOPE (40)	SCOPE (44)	SCOPE (48)	SCOPE (42)	SCOPE (47)
Information security, intrusion detection	SCOPE (57)	SCOPE (55)	SCOPE (60)	SCOPE (59)	SCOPE (52)	SCOPE (62)	SCOPE (53)
Information security, systems, and specifications: hardware	SCOPE (62)	SCOPE (64)	SCOPE (67)	SCOPE (59)	SCOPE (61)	SCOPE (65)	SCOPE (60)
Information security, systems, and specifications: software	SCOPE (62)	SCOPE (55)	SCOPE (60)	SCOPE (59)	SCOPE (61)	SCOPE (69)	SCOPE (67)
Information technology forensic investigation	SCOPE (54)	SCOPE (64)	SCOPE (47)	SCOPE (59)	SCOPE (52)	SCOPE (54)	SCOPE (53)

(Continued)							
	Organization Size (in Terms of Revenue)						
Program	**Most Common**	**< $500M**	**$500M–$1B**	**$1B–$2.5B**	**$2.5B–$7.5B**	**$7.5B–$25B**	**> $25B**
Investigations (e.g. theft, fraud, threats, due diligence, conflict of interest)	BOTH (58)	BOTH (82)	BOTH (73)	BOTH (48)	BOTH (57)	BOTH (58)	BOTH (40)
Law enforcement liaison	BOTH (72)	BOTH (82)	BOTH (67)	BOTH (74)	BOTH (74)	BOTH (65)	BOTH (60)
New product and marketing security	SCOPE (40)	NONE (36)	SCOPE (33)	SCOPE (52)	NONE (39)	SCOPE (42)	SCOPE (47)
Personnel protection/ workplace violence	BOTH (56)	BOTH (64)	BOTH (47)	BOTH (52)	BOTH (52)	BOTH (62)	BOTH (53)
Personnel screening, background investigations, and due diligence	SCOPE (42)	SCOPE (36)	SCOPE (47)	SCOPE (48)	BOTH (26)	SCOPE (58)	SCOPE (40)
Physical site security design and system specifications	BOTH (67)	BOTH (82)	BOTH (67)	BOTH (74)	BOTH (57)	BOTH (65)	BOTH (60)
Product protectio (e.g. tampering, counterfeiting, diversion)	SCOPE (42)	ACCT (27)	SCOPE (33)	SCOPE (52)	NONE (57)	SCOPE (62)	SCOPE (47)
R&D security	NONE (30)	BOTH (45)	BOTH (40)	NONE (37)	NONE (52)	BOTH (27)	SCOPE (33)
Risk assessments	BOTH (63)	BOTH (73)	BOTH (60)	BOTH (67)	BOTH (57)	BOTH (54)	BOTH (80)
Security-related regulations and compliance management	BOTH (77)	BOTH (100)	BOTH (67)	BOTH (85)	BOTH (74)	BOTH (69)	BOTH (73)
Security/business command center	BOTH (55)	BOTH (64)	BOTH (40)	BOTH (59)	BOTH (70)	BOTH (50)	BOTH (53)
Supply chain security	SCOPE (31)	NONE (36)	SCOPE (33)	SCOPE (41)	NONE (39)	SCOPE (42)	SCOPE (40)
Travel security	BOTH (40)	BOTH (27)	NONE (40)	BOTH (37)	BOTH (52)	BOTH (50)	BOTH (53)
Others	NONE (85)	NONE (91)	NONE (87)	NONE (81)	NONE (87)	NONE (85)	NONE (73)

The most common level of responsibility indicated by respondents for a specific security program, categorized by the operating scope of the organization. The percentage (%) of respondents in each category who chose that level of responsibility is included in parentheses.

		Organization Scope			
Program	**Most Common**	**Global/ International**	**National**	**Regional**	**Local**
Asset protection, loss prevention	BOTH (64)	BOTH (67)	BOTH (57)	BOTH (57)	BOTH (50)
Aviation security and operations	NONE (45)	NONE (43)	SCOPE (52)	NONE (71)	SCOPE (50)
Awareness and education	BOTH (58)	BOTH (57)	BOTH (57)	BOTH (71)	RESP (40)
Brand, reputation, trademark/trade name	SCOPE (39)	SCOPE (38)	SCOPE (48)	SCOPE (43)	SCOPE (30)
Business continuity/resiliency	SCOPE (33)	BOTH (32)	SCOPE (33)	SCOPE (43)	SCOPE (40)
Business ethics and compliance	SCOPE (47)	SCOPE (52)	SCOPE (33)	RESP (57)	SCOPE (50)
Business intelligence and analysis	SCOPE (56)	SCOPE (58)	SCOPE (48)	SCOPE (71)	SCOPE (60)
Emergency response and disaster recovery	BOTH (46)	BOTH (43)	BOTH (57)	SCOPE (43)	BOTH (50)
Guard service	NONE (81)	NONE (84)	NONE (86)	BOTH (43)	NONE (70)
Incident reporting	BOTH (71)	BOTH (70)	BOTH (76)	BOTH (71)	BOTH (60)
Information security, application security	SCOPE (54)	SCOPE (54)	SCOPE (71)	BOTH (43)	SCOPE (40)
Information security, intellectual property protection	SCOPE (45)	SCOPE (43)	SCOPE (52)	BOTH (43)	SCOPE (50)
Information security, intrusion detection	SCOPE (57)	SCOPE (54)	SCOPE (71)	BOTH (43)	SCOPE (60)
Information security, systems, and specifications: hardware	SCOPE (62)	SCOPE (63)	SCOPE (67)	BOTH (43)	SCOPE (60)
Information security, systems, and specifications: software	SCOPE (62)	SCOPE (63)	SCOPE (67)	BOTH (43)	SCOPE (60)
Information technology forensic investigation	SCOPE (54)	SCOPE (52)	SCOPE (62)	SCOPE (57)	SCOPE (60)
Investigations (e.g. theft, fraud, threats, due diligence, conflict of interest)	BOTH (58)	BOTH (60)	BOTH (48)	BOTH (71)	BOTH (50)
Law enforcement liaison	BOTH (72)	BOTH (72)	BOTH (62)	BOTH (86)	BOTH (70)
New product and marketing security	SCOPE (40)	SCOPE (41)	SCOPE (43)	NONE (57)	SCOPE (40)

(Continued)

		Organization Scope			
Program	**Most Common**	**Global/ International**	**National**	**Regional**	**Local**
Personnel protection/workplace violence	BOTH (56)	BOTH (59)	BOTH (52)	BOTH (71)	ACCT (30)
Personnel screening, background investigations, and due diligence	SCOPE (42)	SCOPE (43)	SCOPE (38)	SCOPE (57)	SCOPE (40)
Physical site security design and system specifications	BOTH (67)	BOTH (70)	BOTH (67)	BOTH (71)	BOTH (40)
Product protection (e.g. tampering, counterfeiting, diversion)	SCOPE (42)	SCOPE (42)	SCOPE (57)	NONE (57)	SCOPE (40)
R&D security	NONE (30)	BOTH (31)	NONE (38)	NONE (71)	BOTH (30)
Risk assessments	BOTH (63)	BOTH (64)	BOTH (57)	BOTH (71)	BOTH (60)
Security-related regulations and compliance management	BOTH (77)	BOTH (79)	BOTH (67)	BOTH (100)	BOTH (70)
Security/business command center	BOTH (55)	BOTH (58)	BOTH (43)	BOTH (57)	BOTH (60)
Supply chain security	SCOPE (31)	SCOPE (30)	SCOPE (33)	SCOPE (43)	ACCT (30)
Travel security	BOTH (40)	BOTH (56)	SCOPE (38)	RESP (29)	SCOPE (40)
Others	NONE (85)	NONE (88)	NONE (76)	NONE (86)	NONE (70)

The most common level of responsibility indicated by respondents for a specific security program, categorized by the functional area in to which the respondent reports. The percentage (%) of respondents in each category who chose that level of responsibility is included in parentheses.

		Functional Area in to Which the Respondent Reports										
Program	**Most Common**	**Admin.**	**Environmental, Health, and Safety**	**Executive**	**Facilities/ Real Estate**	**Finance**	**Human Resources**	**Information Technology**	**Legal**	**Operations**	**Risk, Audit, Compliance**	**Shared Services**
Asset protection, loss prevention	BOTH (64)	RESP (33)	RESP (50)	BOTH (52)	BOTH (82)	BOTH (63)	BOTH (100)	BOTH (38)	BOTH (73)	BOTH (67)	BOTH (60)	BOTH (100)
Aviation security and operations	NONE (45)	SCOPE (67)	SCOPE (50)	NONE (48)	NONE (53)	SCOPE (38)	ACCT (25)	NONE (54)	NONE (55)	BOTH (67)	NONE (80)	BOTH (38)
Awareness and education	BOTH (58)	ACCT (33)	RESP (50)	BOTH (42)	BOTH (65)	BOTH (50)	BOTH (75)	BOTH (62)	BOTH (82)	ACCT (67)	BOTH (60)	BOTH (38)
Brand, reputation, trademark/ trade name	SCOPE (39)	RESP (33)	SCOPE (100)	RESP (30)	SCOPE (53)	SCOPE (38)	SCOPE (100)	RESP (31)	RESP (36)	ACCT (33)	SCOPE (40)	SCOPE (50)
Business continuity/ resiliency	SCOPE (33)	RESP (67)	SCOPE (50)	RESP (27)	BOTH (53)	BOTH (38)	RESP (75)	RESP (38)	BOTH (36)	BOTH (67)	BOTH (60)	SCOPE (50)
Business ethics and compliance	SCOPE (47)	RESP (67)	RESP (50)	SCOPE (36)	SCOPE (65)	SCOPE (38)	SCOPE (50)	SCOPE (62)	SCOPE (36)	SCOPE (67)	NONE (40)	SCOPE (75)
Business intelligence and analysis	SCOPE (56)	SCOPE (100)	SCOPE (50)	SCOPE (52)	SCOPE (76)	SCOPE (50)	SCOPE (75)	SCOPE (62)	SCOPE (55)	RESP (33)	SCOPE (40)	SCOPE (63)
Emergency response and disaster recovery	BOTH (46)	ACCT (33)	SCOPE (50)	BOTH (52)	BOTH (65)	BOTH (50)	BOTH (75)	SCOPE (46)	BOTH (36)	RESP (33)	BOTH (60)	BOTH (63)

Guard service	NONE (81)	NONE (67)	NONE (100)	NONE (73)	NONE (76)	NONE (100)	NONE (75)	NONE (85)	NONE (73)	NONE (100)	NONE (100)	NONE (75)
Incident reporting	BOTH (71)	BOTH (67)	BOTH (75)	BOTH (67)	BOTH (94)	BOTH (75)	BOTH (100)	RESP (38)	BOTH (73)	BOTH (100)	BOTH (60)	BOTH (50)
Information security, application security	SCOPE (54)	RESP (33)	SCOPE (100)	SCOPE (45)	SCOPE (82)	SCOPE (75)	SCOPE (100)	BOTH (62)	SCOPE (36)	SCOPE (67)	SCOPE (60)	SCOPE (38)
Information security, intellectual property protection	SCOPE (45)	SCOPE (100)	BOTH (50)	RESP (27)	SCOPE (65)	SCOPE (50)	SCOPE (100)	BOTH (46)	SCOPE (36)	SCOPE (67)	SCOPE (60)	SCOPE (38)
Information security, intrusion detection	SCOPE (57)	SCOPE (100)	SCOPE (100)	SCOPE (42)	SCOPE (88)	SCOPE (50)	SCOPE (100)	BOTH (46)	SCOPE (45)	SCOPE (67)	SCOPE (60)	SCOPE (63)
Information security, systems, and specifications: hardware	SCOPE (62)	SCOPE (67)	SCOPE (100)	SCOPE (58)	SCOPE (94)	SCOPE (88)	SCOPE (100)	BOTH (69)	SCOPE (45)	SCOPE (67)	SCOPE (60)	SCOPE (63)
Information security, systems, and specifications: software	SCOPE (62)	SCOPE (67)	SCOPE (100)	SCOPE (52)	SCOPE (94)	SCOPE (88)	SCOPE (100)	BOTH (62)	SCOPE (45)	SCOPE (67)	SCOPE (60)	SCOPE (75)

(Continued)

		Functional Area in to Which the Respondent Reports										
Program	**Most Common**	**Admin.**	**Environmental, Health, and Safety**	**Executive**	**Facilities/ Real Estate**	**Finance**	**Human Resources**	**Information Technology**	**Legal**	**Operations**	**Risk, Audit, Compliance**	**Shared Services**
Information technology forensic investigation	SCOPE (54)	SCOPE (100)	BOTH (50)	SCOPE (48)	SCOPE (88)	SCOPE (75)	SCOPE (100)	BOTH (38)	BOTH (36)	SCOPE (67)	SCOPE (40)	SCOPE (50)
Investigations (e.g. theft, fraud, threats, due diligence, conflict of interest)	BOTH (58)	BOTH (67)	BOTH (75)	BOTH (55)	BOTH (47)	BOTH (63)	BOTH (75)	RESP (38)	BOTH (82)	BOTH (100)	BOTH (40)	BOTH (75)
Law enforcement liaison	BOTH (72)	BOTH (100)	BOTH (75)	BOTH (70)	BOTH (88)	BOTH (63)	BOTH (100)	SCOPE (46)	BOTH (82)	BOTH (100)	BOTH (40)	BOTH (75)
New product and marketing security	SCOPE (40)	SCOPE (67)	SCOPE (50)	SCOPE (33)	SCOPE (59)	ACCT (25)	SCOPE (75)	SCOPE (46)	BOTH (27)	ACCT (33)	NONE (60)	SCOPE (50)
Personnel protection/ workplace violence	BOTH (56)	BOTH (67)	RESP (50)	BOTH (55)	BOTH (65)	BOTH (63)	BOTH (100)	SCOPE (69)	BOTH (73)	BOTH (67)	BOTH (40)	BOTH (63)
Personnel screening, background investigations, and due diligence	SCOPE (42)	RESP (67)	SCOPE (75)	RESP (33)	SCOPE (53)	SCOPE (38)	SCOPE (75)	SCOPE (92)	RESP (27)	ACCT (33)	SCOPE (40)	SCOPE (38)

Physical site security design and system specifications	BOTH (67)	BOTH (100)	BOTH (75)	BOTH (58)	BOTH (88)	BOTH (75)	BOTH (100)	SCOPE (62)	BOTH (91)	BOTH (67)	RESP (40)	BOTH (88)
Product protection (e.g. tampering, counterfeiting, diversion)	SCOPE (42)	RESP (33)	BOTH (50)	SCOPE (36)	SCOPE (53)	SCOPE (38)	SCOPE (75)	SCOPE (77)	NONE (36)	ACCT (33)	NONE (80)	NONE (38)
R&D security	NONE (30)	BOTH (33)	BOTH (75)	SCOPE (27)	BOTH (35)	ACCT (25)	BOTH (50)	BOTH (31)	NONE (45)	BOTH (67)	NONE (80)	NONE (38)
Risk assessments	BOTH (63)	BOTH (67)	RESP (50)	BOTH (55)	BOTH (71)	BOTH (63)	BOTH (75)	BOTH (54)	BOTH (82)	BOTH (100)	RESP (40)	BOTH (75)
Security-related regulations and compliance management	BOTH (77)	BOTH (100)	BOTH (75)	BOTH (73)	BOTH (71)	BOTH (63)	BOTH (100)	BOTH (77)	BOTH (100)	BOTH (100)	RESP (40)	BOTH (88)
Security/ business command center	BOTH (55)	BOTH (100)	BOTH (75)	BOTH (39)	BOTH (71)	BOTH (50)	BOTH (100)	BOTH (46)	BOTH (64)	BOTH (100)	RESP (40)	BOTH (75)
Supply chain security	SCOPE (31)	RESP (33)	ACCT (25)	SCOPE (30)	SCOPE (35)	RESP (38)	SCOPE (75)	SCOPE (62)	BOTH (55)	BOTH (67)	NONE (40)	NONE (38)
Travel security	BOTH (40)	RESP (33)	BOTH (75)	BOTH (27)	BOTH (35)	BOTH (75)	BOTH (50)	SCOPE (54)	BOTH (91)	BOTH (100)	SCOPE (60)	BOTH (63)
Others	NONE (85)	NONE (100)	NONE (100)	NONE (82)	NONE (88)	NONE (88)	NONE (75)	NONE (62)	NONE (91)	NONE (67)	NONE (100)	NONE (100)

APPENDIX A

Board-Level Risk Categories

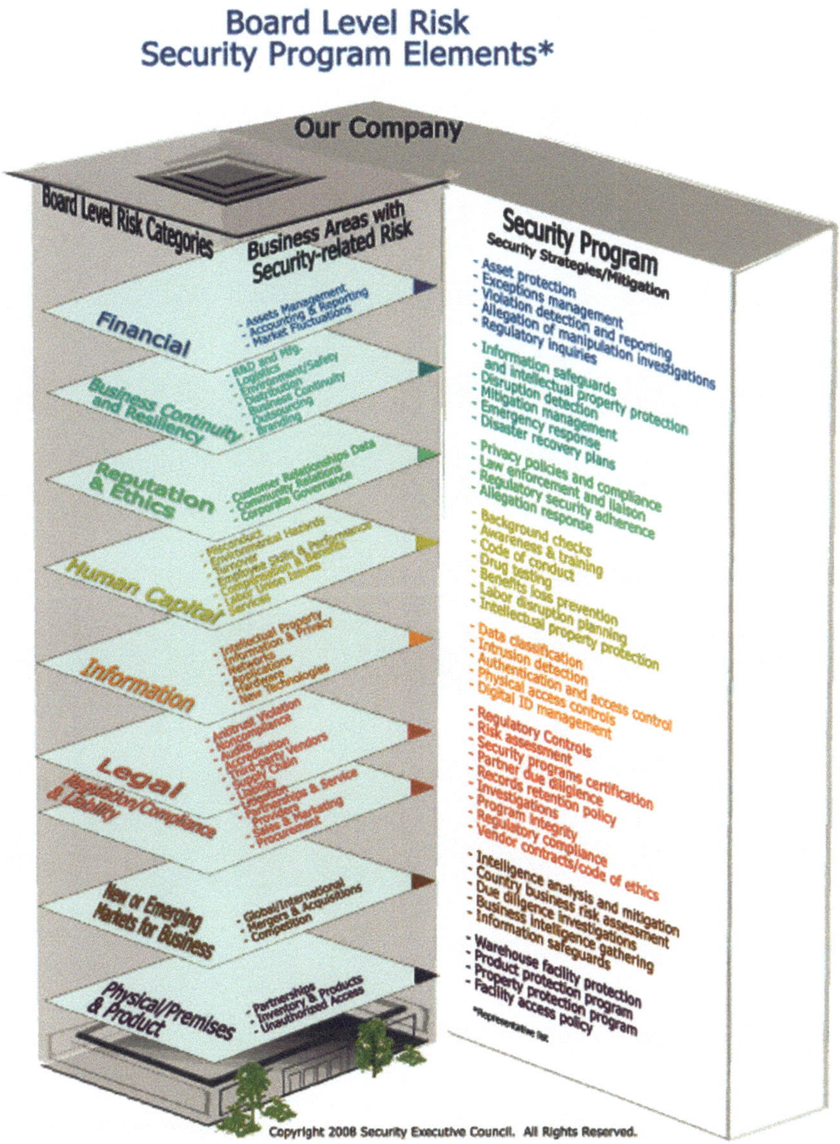

For more on what this chart represents and why it is important, see "Managing Enterprise-Wide Board Risk," from the Security Executive Council: https://www.securityexecutivecouncil.com/spotlight/?sid=25915.

APPENDIX B

Data Cleansing

In any research, steps must be taken to ensure the integrity of the resulting analysis. In order to produce the most valuable report possible, the originally inputted data was examined to ensure that any obviously unusable or unreliable information was removed. Details regarding this data cleansing are given below.

Not all respondents answered all questions. Respondents that provided no data or minimal valuable data were removed from the analysis. In the few cases where a respondent may have taken the survey more than once, the duplicate entry was removed. If multiple people from the same organization took part in the survey, a decision was made regarding whether they represented sufficiently distinct areas of the organization. If this could not be determined the respondent representing the more senior position in the organization hierarchy was retained and the other response was removed.

When appropriate and meaningful, responses to questions with an "other" option were corrected to match one of the given response options provided in the survey question.

In some select cases, when too few respondents existed in the original categories, data was consolidated to reduce the number of variables and/or categories. This was necessary to make the resulting analysis more meaningful and to ensure anonymity of the participants.

For selected data elements, actual publicly reported values were used in place of the data ranges in the original survey. In particular this was done for revenue and employee counts when actual numbers were more meaningful than the ranges used in the original survey.

Because not all respondents answered every question, extra effort was made to ensure the integrity of the analysis for each data element used in crosstabs. For example: Only respondents that provided a complete security budget as well as organization size and revenue figures were used in the analysis of the security budget section of this research. If a respondent did not provide sufficient information to meet the needs for any of the security budget analyses, they were not included in any of the security budget analyses.

It became clear during the analysis that a couple of institutions performed unique activities that resulted in the relative sizes of their security budgets being outliers to the norm. These organizations were excluded from the security budget analyses to avoid the otherwise resultant skewing of the data.

REFERENCE

Hayes, B., Kotwica, K., Lefler, D., 2010. March 1. Managing Enterprise-Wide Board Risk. Corporate Secretary. Retrieved from <http://www.corporatesecretary.com/articles/regulation-and-legal/11166/managing-enterprise-wide-board-risk/>.

ABOUT THE AUTHORS

Bob Hayes has more than 25 years of experience developing security programs and providing security services for corporations, including eight years as the CSO at Georgia Pacific and nine years as security operations manager at 3M. His security experience spans the manufacturing, distribution, research and development, and consumer products industries as well as national critical infrastructure organizations. Additionally, he has more than 10 years of successful law enforcement and training experience in Florida and Michigan. Bob is a recognized innovator in the security field and was named as one of the 25 Most Influential People in the Security Industry by *Security Magazine*. He is a frequent speaker at key industry events. He is a leading expert on security issues and has been quoted by such major media outlets as the *Wall Street Journal* and *Forbes*. Bob is currently the managing director of the Security Executive Council.

Greg Kane has held a director role for the Security Executive Council for more than seven years. In this role he is responsible for mitigating risk as it applies to IT systems and the extensive intellectual property assets contained within. He has been responsible for disaster recovery and business continuity for various organizations for over 20 years. His work experience also includes analysis of security-related regulations, standards, and guidelines in order to encourage efficient and value-added compliance management. Greg leverages his strong skills in research and analysis to write a monthly security newsletter published to an audience of over 10,000 security practitioners. Before joining his current employer, Greg provided services to multiple businesses from retail to high tech manufacturing. This included over 10 successful years with a leading international business consulting services provider. Greg's educational background includes an MS degree in computer science and an MBA.

Kathleen Kotwica has a PhD in experimental psychology from DePaul University and has had a career as a researcher and knowledge strategist. Her experience includes positions as information architecture consultant at a New England consulting firm, director of online research at CXO Media (IDG), and research associate at Children's

Hospital in Boston. She has authored and edited security industry trade and business articles and spoken at security-related conferences including CSO Perspectives, SecureWorld Expo, ASIS, and CSCMP. In her current role as EVP and chief knowledge strategist at the Security Executive Council she leads the development and production of Council tools, solutions, and publications. She additionally conducts industry research and analysis to improve security and risk management practices.

About Elsevier's Security Executive Council Risk Management Portfolio

Elsevier's Security Executive Council Risk Management Portfolio is the voice of the security leader. It equips executives, practitioners, and educators with research-based, proven information and practical solutions for successful security and risk management programs. This portfolio covers topics in the areas of risk mitigation and assessment, ideation and implementation, and professional development. It brings trusted operational research, risk management advice, tactics, and tools to business professionals. Previously available only to the Security Executive Council community, this content—covering corporate security, enterprise crisis management, global IT security, and more—provides real-world solutions and "how-to" applications. This portfolio enables business and security executives, security practitioners, and educators to implement new physical and digital risk management strategies and build successful security and risk management programs.

Elsevier's Security Executive Council Risk Management Portfolio is a key part of the *Elsevier Risk Management and Security Collection*. The collection provides a complete portfolio of titles for the business executive, practitioner, and educator by bringing together the best imprints in risk management, security leadership, digital forensics, IT security, physical security, homeland security, and emergency management: Syngress, which provides cutting-edge computer and information security material; Butterworth Heinemann, the premier security, risk management, homeland security, and disaster-preparedness publisher; and Anderson Publishing, a leader in criminal justice publishing for more than 40 years. These imprints, along with the addition of Security Executive Council content, bring the work of highly regarded authors into one prestigious, complete collection.

The Security Executive Council (www.securityexecutivecouncil.com) is a leading problem-solving research and services organization focused on helping businesses build value while improving their ability to effectively manage and mitigate risk. Drawing on the collective knowledge of a large community of successful security practitioners, experts, and strategic alliance partners, the Council develops strategy and insight and identifies proven practices that cannot be found anywhere else. Their research, services, and tools are focused on protecting people, brand, information, physical assets, and the bottom line.

Elsevier (www.elsevier.com) is an international multimedia publishing company that provides world-class information and innovative solutions tools. It is part of Reed Elsevier, a world-leading provider of professional information solutions in the science, medical, risk, legal, and business sectors.

Kolibri 360 GmbH
Ferdinand-Jühlke-Straße 7
99095 Erfurt, Deutschland
produktsicherheit@kolibri360.de